Clough's Bantam Book

Devoted Exclusively to Bantam Chickens – How To Mate, Breed and Care For Them

by W.W. Clough

with an introduction by Jackson Chambers

This work contains material that was originally published in 1896.

This publication is within the Public Domain.

*This edition is reprinted for educational purposes
and in accordance with all applicable Federal Laws.*

Self Reliance Books

Get more historic titles on animal and stock breeding, gardening and old fashioned skills by visiting us at:

http://selfreliancebooks.blogspot.com/

Introduction

I am pleased to present yet another title on Poultry.

The work is in the Public Domain and is re-printed here in accordance with Federal Laws.

As with all reprinted books of this age that are intended to perfectly reproduce the original edition, considerable pains and effort had to be undertaken to correct fading and sometimes outright damage to existing proofs of this title. At times, this task is quite monumental, requiring an almost total "rebuilding" of some pages from digital proofs of multiple copies. Despite this, imperfections still sometimes exist in the final proof and may detract from the visual appearance of the text.

I hope you enjoy reading this book as much as I enjoyed making it available to readers again.

Jackson Chambers

BANTAMS.

BUFF COCHIN BANTAMS.

For description see page 17.

DRAWN BY
Louis P Graham
1896

DRAWN FOR
CLOUGH'S
BANTAM
BOOK

Yours Respectfully,

W. W. Clough.

INTRODUCTION.

Having received many letters from amateurs and others asking for a description of this or that variety of Bantams, and as there is at present no reliable book, at a reasonable price, that treats on Bantams exclusively, the author being assured by many prominent Bantam breeders that such a book was badly needed, decided to publish this work. It will be the aim of the author not to puff his own stock, or advise beginners to buy of him, or to in any way use its pages, except to impart information and to give as near as possible the exact requirements of the different varieties, including the bad as well as the good points.

Bantams of all varieties are receiving more attention at present than they have in the past, and from indications a boom is near at hand which all lovers of the little creatures will be glad to welcome.

Yours for Bants,

W. W. CLOUGH.

CONTENTS.

BANTAMS.

———o———

How to Mate, Breed and Care for them for Pleasure and Profit.

A BANTAM is supposed to be the exact reproduction on a very small scale, of the varieties from which they are descended. A Cochin Bantam is supposed to be a little Cochin; a Rose Comb Bantam a little Hamburg; a Game Bantam a diminitive Game. While they are supposed to be like the large varieties from which they have descended, still they are not in all respects. The standard requirements are in some breeds slightly different in the Bantams than in the large varieties. These differences will be explained under the description of the various breeds and varieties.

The latest Standard of Perfection recognizes 14 varieties of Bantams besides the Game. These include the Silver and Golden Sebrights; Buff, Black, White and Partridge Cochin; Booted White; three varieties of Japanese; Rose Comb, Black and White; Plain and Bearded White Crested White Polish. There are nearly as many more varieties which are not yet recognized. Some will probably be admitted at the next revision of the Standard, while it will be years before others get in, if ever. The Light and Dark Brahma have a fair start. Mr. Zimmer has originated the Buff Polish and is now at work on the Golden Bearded variety. The Buff Polish has quite a start, and when through scientific breeding a fair per cent of good even marked birds can be relied upon they will, in the writer's opinion, be one of our leading varieties of Bantams. Rumples and Silkies, while not classed among Bantam varieties, could be justly placed among the list, they being as small as many of the Bantams. It is wonderful how the little Bantams have gained in popularity during the past few years.   Pheasant breeders use the Bantam for setting, prefering them to any large fowl.   I have sold very many Buff Cochin hens for this purpose.   The Cochin varieties are considered superior as setters and mothers to any other variety of Bantams.

The profit derived from a flock of Bantams is more than any person not familiar with the little creatures would imagine, taking into consideration the time it requires for them to come to maturity, the small amount of food they consume, and the space necessary to accommodate the flock.   A Cochin Bantam will lay as many eggs in a year as a Cochin or Brahma, a Sebright Bantam will lay as many as a Leghorn, Minorca or Hamburg. Twice the number can be kept, without crowding, in the same space.   They will consume only about one-third the amount of food, and will lay an egg about three-fifths the size of a large fowl.   Taking everything into consideration they will be found to yield a better profit than the large varieties.   Of course their eggs are not salable at full prices, but can readily be marketed at over one-half the going price.   The only drawback is the poultry, they are not salable on account of their size, still they are extremly fine eating, and by killing two at a time will furnish a meal for a good sized family.

Mr. H. S. Babcock, who has bred Bantams for many years, writes as follows: "Bantams are peculiarly adapted for pets. Their small size is of great advantage in this respect.   A pet bird is one that should be able to be handled easily.   A large fowl, simply because it is large, can not be easily handled, but a Bantam weighing a pound or a pound and a half can be held on the outstretched hand without weariness.   The small size is of great advantage, therefore, not only in appealing to our affections, but in enabling the affections to be manifested without any weariness.

Bantams are easily tamed.   This seems to be true not only of the fearless little Games but of the very domestic little Cochins, and of all classes intermediate between these extremes.   A pet bird should be, must be, one that can be rendered tame without difficulty, and Bantams, therefore, meet this very obvious requirement.

A pet bird, also, ought to be a beautiful bird. And Bantams are beautiful. They have all the rich coloring, exquisite and accurate markings, and beautiful and graceful shapes which make all domesticated poultry so attractive. And all these elements seem to be especially refined by the small size. We wonder that a little Bantam can possess all these qualities and therefore find our admiration sensibly increased.

Pets serve an important purpose in the economy of life. They meet a demand of human nature. They seem to interest and instruct youth and afford rest and recreation to the eye. The heart of man does not seem to be wholly satisfied by the love of his fellow man. He turns to floriculture and rejoices in the beauty of the blossoms his art has created. But flowers leave something to be desired. They can not show that they are appreciated. They can not come to meet one and can not turn upon him an intelligent eye as he caresses them. But a Bantam can. They show their appreciation of man's care and affection. They follow his footsteps with intelligent devotion. They rejoice his heart, not only with their great beauty but with the manifestations of their delight in his presence. They are as lovely as the lovliest blooms of earth, and they are alive. Living, breathing, acting, responsive, intelligent flowers they are, ministering to man's sense of beauty. And not only this but also do they afford a great fund of amusement and interest by their amusing ways, their sense of pride, their domestic relations. In their daily search for food, their courtships, their gallantness, their challenges, their battles, their nesting operations, their brooding of the young and the like, they afford a study of absorbing interest. Nor are they to be despised even by the scientist who can trace the rudiments of a language in their varied sounds, can follow back a habit to its original in the wild state, and can obtain from their study great light upon biological problems. Pet Bantams deservedly occupy a prominent place in life and are worthy not only of the place they now occupy but of a much higher and much larger place. And it is probable that the place they deserve will, after a time, be bestowed upon them, and their numbers as pets be greatly multiplied."

HATCHING AND BREEDING.

When to hatch Bantams is a matter of some importance to the Bantam breeder. The common custom of hatching late in the season so as to have the birds stunted in growth has in its favor the small size attained, and if applied only to the short-legged varieties, such as Cochin and Japanese, has little objection save the possible sickness that may come to immature birds. One who practices very late hatching ought to have specially good accommodations for young birds to prevent sickness, suffering and loss. We prefer to get our Bantams small by selection rather than by dwarfing and so we prefer the months of April, May and June to any others. July and August we do not like on account of the excessive heat. September is pretty late and October and November we dislike because of the cold to which the young birds will be subjected.

How to hatch is not so important, for either a good hen or a good incubator will do the work admirably. We have used both and have been pleased with the results, finding little to choose between them, save the fact that we like the artificial method the better as being the least troublesome.

In rearing we feed liberally at first to give the birds a good start, afterwards gradually reducing the amount of the ration. But we never starve our Bantams. We prefer to take the chance of a few becoming over weight and have them to eat, than to inflict unnecessary cruelty upon the young.

Dry feed is our hobby. Those who prefer wet-up messes are welcome to do so, but we can raise a much larger percentage on dry than we can upon wet-up food.

Our feeding consists of millet seed, bread crumbs, boiled eggs, finely sifted cracked corn, Irish oatmeal, and wheat as soon as the chicks are large enough to eat it. We feed also considerable ground beef scraps. This is our method. We give to a brood—if in a brooder of from 40 to 50—one or two hard-boiled eggs, cut in halves, to pick at. We mix for the earlier feeds—first two or three weeks—about equal parts of cracked corn, millet seed, oatmeal and beef scraps. Later we remove the millet seed and substitute whole wheat.

Fine grit we keep constantly before the chickens. We have been using the prepared Mica Grit and like it. Other kinds, if of the right size, may be equally good, but the Mica answers nicely.

We keep water before the chickens, in fountains, at all times. The fountains are carefully cleaned each day, and if dirt is scratched into the lip during the day this is cleaned out and the fountain rinsed.

Green food is given—fresh grass, lettuce and the like, as we may be able to furnish it. We do not allow the young chicks to run in the wet grass until they are old enough to be weaned and require no warmth other than that furnished by their own bodies. Then they are turned into a field of young clover, which we have sown every year, and are let out of their coops every morning about seven or seven and one-half o'clock, irrespective of whether there is dew on the grass or not.

By this method we have had very good success in rearing Bantams and very few birds have been over weight. We do not know of a better method; if we did we would adopt it.

GOLDEN SEBRIGHT BANTAM HEN.

The Sebright Bantams.

IN taking up the different varieties of Bantams the Sebrights are the first to be considered, not simply because they are the first mentioned in the Standard, but because their origin dates back over a century and are one of the oldest and probably best known varieties.

The Golden and Silver are both the same size, shape and in every particular marked the same, except the feathers in the Golden variety are a rich golden color laced with black and white, the feathers of the Silver are a pure white with black lacing. Both the Golden and Silver have rose combs, horn-colored beaks, legs a slaty blue, while the ear lobes can either be red or white, or a mixture of the two. Now here I think the new Standard took a long step backward, for in speaking of the ear lobes they say "color immaterial", so that now we will not be able to improve on them for a while, as it does not make any difference what color they are.

GOLDEN SEBRIGHT BANTAMS.

The old Standard calls for white ear lobes, and most of the hens and pullets have more or less white, but we notice the cocks and cockerels almost always have solid red ear lobes, and for that reason I think it would be easier to breed out the white and in time they would be all solid red. I like to see a solid colored ear lobe.

The Sebrights present the finest examples of success, with what might be called artificial breeding, in the world. They are entirely made up from crossing various varieties together, and yet they breed most remarkably true. The black lacing around their feathers, and their low set rose comb are hard to breed into a fowl, as experiments of other fanciers have proven. Not only here are they remarkable, but the males are entirely hen feathered, having no hackles, saddle feathers or sickles—the only instance of such a combination of omissions in all the fowl kingdom.

Sir John Sebright, from whom they take their name, has the honor of having originated them, and it is said that he was over twenty years in bringing them to perfection. They are good layers of white eggs, and will lay probably as many in a year as a Leghorn, Minorca or Hamburg. They are good setters and mothers. The young are easily raised by taking care not to allow them to get wet, or feeding soft food that is liable to sour in their crops, until they are two or three weeks old. The eggs do not, as a rule, hatch as well as from some other varieties.

For the Sebright feathers shown in this book and the comments thereon, we are indebted to Judge B. F. Zimmer, of Gloversville, N. Y. He says: "No. 1 is a good hackle feather. In this section the light undercolor is an advantage as the hackle is required to be frosted outside with silver or gold, according as they may be Gold or Silver Sebrights. Nos. 2 and 3, although far above the average breast feathers, are still both slightly defective, No. 2 showing a *very slight* tinge of frosting outside the black lacing half way down toward the fluffy part of the feather. The only apparent objection to No. 3 would be to say the lacing was a trifle wide; however, a bird clothed with feathers of this stamp in all sections would be beyond value as a breeding bird, and would cause any judge to score them close up to the 100 mark, and yet by comparing them you will readily see No. 1 excels in shape of feather and shape and size of white center.

No. 4, tail covert, perfection. No. 5, covert, good, slightly defective in shape of end of feather and speck of black running in the white center, near fluffy part of feather. No. 6, feather from the center of back, a good one.

No. 7, center of back. Although a bird with such feathers looks very handsome, still from a judge's, a scientific breeder, or a critic's standpoint as to perfection it must be considered imperfect, as lacing is irregular in width, color of lacing up side of feather is dull, caused by defective undercolor which is too light to allow of intense outside color."

Mr. Zimmer says that, "one must not think the woods are full of birds with better feathers than these." We notice, quite often, many specimens where the lacing on the sides of the feather are not only narrow, like No. 7, but the black edge will be entirely

Golden and Silver Sebright Bantams.

er in any section that is laced correctly. It may be narrow up the side of feather, and on the end is apt to be quite wide. Again, it may be altogether too wide around the entire feather. The latter is less faulty than the former, if those too widely laced are evenly laced up the side and around the end, 1-16 of an inch being about the correct width of lacing in either variety, and black (not dull, brownish black or gray) is the correct color of the lacing. And bear in mind that to get this strong outside color you must have under color either dark slate or black. Very many specimens that from outward appearance look very pretty and perfect, show, when opened up, but few feathers laced clear round, the black not meeting at the shaft where the downy part begins.

Again, many lose sight of the Sebright shape entirely when striving for lacing. The short back, well-spread fan tail, and the large, drooping wings are forgotten. We personally know of men who have bought good stock, at good prices, who in five years had only fair birds from them. How to mate and breed your birds cannot be told on paper, farther than to say, choose strong, vigorous breeding stock, such as have more than one or two strong points, and use judgement in remedying defects; use birds of known good strains; don't be afraid to in-breed; anybody worthy the name of scientific breeder knows how to in-breed. Never in-breed unless you desire to intensify a quality, if you are a novice, and then breed, after first cross, a sire to a daughter or dam to a son. We have seen where a sire was bred to daughters of succeeding years for seven generations with decidedly good results; have seen desired results obtained in two seasons. If you don't know what you are doing, and what for, better not meddle with in-breeding. Hatch in May, June, July and August.

As regards food, use nothing wet, and after three days old, only an occasional meal of anything moist. Spratt's food dry, oat meal, cracked wheat and corn. If chicks are in yards, cut grass, and green bone fresh, a light feed each day. Don't kill them by feeding oftener than four times daily.

Following these suggestions, you ought to be able to raise the healthiest and handsomest of Sebrights. In relation to in-breeding and how far to go, it all depends upon the improvements made.

Rose-Combed White Bantams.

JUDGE ZIMMER says that, "good Rose-Combed White Bantams are seldom seen at our leading shows." A good specimen is a rarity, not because they are a hard variety to breed, but simply because they have not had the attention from breeders that they deserve. For the last year or two I have noticed more of them than formerly and the interest in them seems to be increasing. They are very stylish in appearance, good layers, setters and mothers. They lay a white egg, and as many of them as most any Bantam variety. They are as small as any variety, the standard weight being the same as the Sebrights. In color they are a pure snow white, and any other color or shade would debar them from the exhibition room. The face and wattles should be a bright red. The beak should be slightly curved, rather short and yellow or white in color. The comb, rose, well covered with small points, rather small, and in color bright red. The ear lobes should be a pure white. The tail of the male should be well expanded and carried upright, as shown in cut. The breast should be full and round. The legs and toes should be rather short and in color yellow or white, if white they should have a pinkish or flesh-colored tinge. Should a specimen have any feathers on its legs or toes, its comb be other than a rose comb, or should there be a natural absence of spike, the bird would be disqualified.

The time is not far distant, I believe, when the Rose-Combed White will be more of a favorite among breeders than at the present time.

Rose-Combed Black.

A ROSE-COMBED Bantam is supposed to be a diminitive Hamburg. The principal difference between the White and Black is the color. The Black variety is the best known, and are more often seen at our shows. The Rose-Combed are sometimes called the Black or White African Bantams. The illustration of the White variety answers also for the Black, they being the same style, shape and size.

In breeding this variety select birds that have a good lustrious black or greenish black plumage, avoiding the dull, rusty or purple shades. The legs should be a very dark, leaden blue or black. The beak a very dark horn or black. In all other points the Black variety are like the White.

Booted White Bantams.

THIS variety is not as popular as many, but still has its admirers. They are very showy and pretty, and deserve much more attention than they are at present receiving.

. The chief point in this variety that breeders aim for is to have them very heavily booted. The thighs should be of medium size, with long, stiff feathers known as "vulture hocks," the heavier the better. They should extend so as to nearly touch the ground. I have seen specimens where these hock feathers actually did drag on the ground when the bird was in an erect position. The shanks and toes must be either white or yellow in color, white prefered, both the shanks and toes to be heavily feathered. The beak should correspond in color with the legs, comb single, ear lobes and wattles a bright red. The tail should be very upright, with long, well curved sickle feathers. The plumage throughout must be a pure white; should there be any other color feathers in a specimen it would be disqualified in the show room.

Buff Cochin Bantams.

THE Buff Cochin or Pekin Bantam I think ranks first in popularity, not only first among the Cochin varieties but among all varieties of Bantams. More Buffs are seen at our shows than any other one variety. I receive, surely, two orders for this beautiful breed to one of any other. Buff seems to be the popular color, not only in Bantams but in large breeds. If you look through the advertisements of any poultry paper you will find headings that read, "Buff Crank," "Buffs Only," etc. This shows the great interest taken in the different Buff breeds. What is prettier than a nice, even, rich buff color fowl. I think few breeders think any other solid color excels it. The Buff Bantams have been scientifically bred and come truer to Standard requirement than any of the other varieties of Cochin Bantams. The Standard weight is two ounces less and still they are not cut on weight at our shows any more than the Black, White or Partridge varieties.

The history of the Cochin or Pekin, as far as known, shows that the first birds obtained in England, came from the city of Pekin, China, when they were obtained from the yards about the emperor's palace, during the sacking of that place by the French soldiers during the Franco-Chinese war of 1860.

The great aim of the breeder of Buffs should be to get a clear, rich buff color throughout. There is no standard shade, and many judges differ as to the proper shade. The writer thinks it of less importance to have a particular shade than to have the bird a clear buff. A perfectly clear buff that is entirely free from white, black or that mealy appearance so often noticed is a rare thing. A bird may look even in color while in the pen, but when taken in the hand and its wings, tail, and especially its under color examined, we find the clear colored specimens are not in the majority. Under color is as much importance as the surface color; this, remember, holds good not only in this variety but in all other varieties, whether they be Bantams or large fowls. The under color is generally lighter than the surface color, but there is a vast difference between a light shade of buff and a white under color.

Judges do no pay enough attention to the under color. In breeding I believe the very light under color birds are apt to bring out chicks that will show more or less white when matured. If the under color be very dark the offspring from such birds are liable to show more or less black. Should a very light male be mated to light females the result will be white feathers in the wings of many specimens. If the male and female be very dark then the result will be vice versa. The cock or cockerel is much darker than the female in this variety, and should be a deep buff or reddish orange, avoiding the clear red as much as possible; the color should be uniform throughout, except the tail which should be a dark chestnut. The shape is important and should be as shown in cut; head, small; beak, yellow; eyes, bright bay; comb, single, and rather small and evenly seriated; ear lobes, red; neck and back of males should have abundant hackle and saddle feathers flowing well over the shoulders and saddle; breast, broad and full; fluff, very abundant; tail should be carried rather horizontally, broad but short. Legs and toes, this is a very important section. Several years ago this variety was bred with greenish or willow color legs; in the present Standard the color is changed to yellow, which is the proper color. Many dark leg specimens still appear, even in the very best of flocks. I have also noticed several birds with whitish color legs. Either of these will disqualify. The color must be yellow. Do not breed from the dark leg birds, the writer has had experience enough from breeding one season from a few dark leg pullets that were otherwise fine bred, to last him a life time. The effect of such breeding will last many generations. The best birds are never too good for breeding, and a disqualified bird is rarely too good for soup.

The feathering on the shanks should be heavy. The toes should be feathered to their extremity, both the outer and middle toes. Five toes are sometimes seen in this variety which was the old style of breeding, and still crops out, occasionally, like the willow colored legs. There should be but four toes, five disqualifies the specimen. It is not uncommon to see, at our small shows, birds with either green legs or five toes. I noticed some at a Providence show, which, by the way, was not a small one, with dark legs, also at a Medway show with five toes; of course both lots were disqualified and not allowed to compete, but it only goes to show that there are plenty of these old style birds still in the hands of inexperienced breeders.

Buff Bantams are good layers, the best of setters, and unexcelled as mothers. The Poultry World says, "We know of a Pekin Bantam hen which laid over sixty eggs in one season, besides hatching and rearing a brood of chickens, aad her eggs were almost large enough for marketing; yet the food she consumed was so small in quantity as to show a handsome profit for her keeping."

Partridge Cochin Bantams.

THIS variety is very pretty when well bred, but there are but few fine specimens to be found. The shows are few that have them, and when you do see them on exhibition they are, as a rule, too large, and resemble the large Partridge Cochin more than a pigmy Bantam; still I have seen good specimens that were down to Standard weight.

The author is sorry not to be able to have an illustration of this variety. The general shape and color is the same as the large Partridge Cochin. The male has a striking contrast in plumage. The head is red; neck, red or reddish orange, with a distinct black stripe in each feather of the hackle, which should be very abundant and flow well over the shoulders; plumage of the back, same as the the neck, with saddle feathers striped like those of the hackel; tail, greenish black; breast, a rich, very deep black. The plumage of the female is a reddish brown penciled with dark brown, the specimen to be evenly penciled.

The Partridge Cochin Bantam is a difficult variety to breed true to Standard requirements, and to produce nice exhibition birds requires two matings, one for males and another for females; other fair results may be secured by one mating.

Drawn By
Louis P. Graham
1896
For
CLOUGH'S BANTAM BOOK

WHITE COCHIN BANTAMS.

White Cochin Bantams.

TO my fancy the White Cochin is excelled in beauty only by the Buffs. Of course we all have our favorite color, but what is really much prettier, when kept perfectly clean, than a pure snow white Bantam. The White Cochin is surely gaining in popularity as fast as any of the little creatures, and rank as second among the Cochin varieties. They breed fairly true to color, and are well down in size. Very few of our best specimens shown are cut in weight at our large shows, but at the country fairs are apt to overrun somewhat. Birds are not generally weighed at the fairs, so they pass unnoticed; in fact birds are not satisfactorily judged at many fairs. I have now in my yards a White Cochin cockerel that took first premium at a New York fair that is disqualified, having willow color legs; the judge in this case, as in many others, didn't know his business.

White Cochin Bantams should be pure white throughout, the whiter the better. It is not easy to get all specimens pure snow white, especially in the male. They are apt to show a creamy or straw color tint. This does not disqualify a specimen unless it extends below the surface, but such a bird would be cut from one to two points on color. Comb must be single and even; beak, yellow; eyes, bay; ear lobes and wattles, a bright red; hackle feathers should be abundant, as also should the saddle feathers; tail, carried rather horizontally; shanks and toes, yellow or white. While either color is admitted by the Standard, **there is no question but the legs should be yellow**. No breeder of prominence will dispute this, and I should strongly advise breeders only to breed from yellow leg birds. The greenish tint is quite apt to show in some strains of White Cochins, but bear in mind such birds are disqualified, and from a fancier's standpoint are worthless. The legs and feet must be feathered heavily, the heavier the better; there is no danger of getting any of the Cochin varieties too heavily feathered. What is termed as "vulture hock," and which is a disqualification in all large Asiatic fowls, is considered essential in the Cochin Bantams of all varieties. They are not disqualified without the hock feathers, but a bird with them is generally considered superior to one without them.

DRAWN FOR
CLOUGH'S
BANTAM
BOOK

Louis P. Graham
1896

BLACK COCHIN BANTAMS.

As layers, setters and mothers there is no difference in the Cochin varieties. The Cochins cannot be excelled as setters and mothers, as layers they are on a par with the large Asiatics.

Black Cochin Bantams.

THE Black Cochin Bantam is as handsome as a black fowl can be. I do not mean by this that a black fowl is not handsome, because I think a nice, bright, greenish black fowl that is kept clean and free from dust, is a beautiful sight. In the Black variety of Cochin Bantams many of our ablest judges and breeders are deeply interested, and within the last three or four years have made vast improvement in this as in all varieties of Bantams. There is still plenty of room for improvement, especially in the male bird. The Standard disqualifies for white in any part of the plumage if it extends over one-half inch in any single feather, or if two or more feathers are tipped or edged with positive white; this applies to any portion of the bird except toe feathering, in which white does not disqualify. It is nearly impossible to breed a male bird that will be clear black to the skin, the under color in the hackle and at the base of the tail will be white or grayish white in probably 90% of all the cockerels, and 95%, if not more, in all cocks. Judge Zimmer says, " a cock of this variety that is black to the skin, and first-class in other respects, is worth $50, and a cockerel $25." The cockerel that is clear black to the skin will after shedding show some white in under color in probably nine cases out of ten. I have a cock in my yards now that was nearly black under color when a cockerel, but is now very white under hackle and saddle, so much so that he would be disqualified. To illustrate how extremely scarce the clear black under color male birds are it is only necessary to say that at the Boston show in 1896, which was the grandest show ever held in this country, there was not a Black Cochin Bantam cock on exhibition that was good enough in color to receive a first prize, and only second prize was awarded to the best bird.

BLACK TAILED JAPANESE BANTAMS.

The females breed truer to color, while light and sometimes white under color will show in good bred flocks, still they breed much truer to color than the males. I have dwelt on color more in this variety than in others because it is the all-essential point that breeders must aim at. In time, by careful mating, the male can be bred so as to come nearer Standard requirements as to under color. Red is often seen in the hackle or saddle of the male, but seldom in the female; some breeders claim that a male with a reddish tint will produce better pullets than a clear black male, but I prefer to breed from as clear a black male as can be obtained.

The shanks in this variety must be either black or black with a slightly greenish tint, and bottom of the feet yellow. The general shape is like that of all the Cochin Bantams, A good illustration of the leading Cochins are shown in this book and represent the true Cochin shape as near as can be drawn by an artist.

Black Tailed Japanese Bantams.

OF the three varieties of Japanese Bantams the Black Tailed is supposed to be the original. They are not very common and require great skill in breeding to produce fine specimens. If a breeder can get 25% good marked birds it is considered good for this variety.

Japanese Bantams are the most odd of any of the varieties, being very short in their legs and carrying their wings drooping almost to the ground. Their heads are very small and finely shaped; their combs being high and single. The color is white, except wing flight and tail, which are black; the latter having very long sickles, which in good specimens are laced on the outer edge with white. They are very coquettish in manner and make delightful pets. Japanese eggs are white and the smallest in size, on an average, of any of the Bantam eggs.

In mating select the best specimens and breed only from these. The plumage should be a pure white throughout, except tail and wings. The wings, when folded, should show only white, but when spread, if the specimen be a good one, will show primaries to be a dark slate or black, edged with white; secondaries, dark slate or black, with a wide edging of white on the upper web, the lower web white; coverts, white. The tail is an important and conspicuous portion and must be very large and carried, in the male, very erect, so that the back of the head and tail will nearly touch; color, black; the sickles very upright, somewhat curved, good length and of a rich, glossy black edged, with white; the coverts are of the same color as the sickles; legs and feet a bright yellow. Many specimens, I notice, have a whitish or dull yellow color legs, but such birds are disqualified for show purposes, neither should they be used as breeders. The face, eyes, comb, wattles, and ear lobes should be bright red; beak, yellow; hackle and saddle feathers abundant. Good specimens of this variety can be sold at high prices.

White Japanese Bantams.

THE general shape of the White Japanese is like that of the Black Tailed, and the description of the latter can be applied in all respects to the White, except the plumage, which is white throughout. The White variety is preferred by many; they being a solid color will breed truer to feather. Any white breed of fowls will show, now and then, dark feathers in the plumage, but feathers of any other color than white disqualifies the White Japanese.

Black Japanese.

THE Black Japanese also differ but very little from the Black Tailed, except in color of plumage which must be a lustrous black throughout in both male and female. The eyes are a darker red than in the Black Tailed, while the shanks may be the same, which is yellow, or it may be shaded with black. With these few exceptions the standard is the same.   Birds that have shanks other than yellow or yellowish black, or pure white in any part of the plumage that extends over one-half inch, or if two or more feathers are tipped with pure white, such birds are barred from the show room as disqualified specimens.   The Black Japanese are not a common variety, but those who breed them speak highly in their favor.

BEARDED WHITE POLISH BANTAMS.
BRED AND OWNED BY FRANK HURST, WATERLOO, IOWA.

Polish Bantams.

THE White Crested White is the only variety of Polish Bantams that are recognized by the present Standard.   There are severel other varieties bred, though not yet perfected; these include the Silver, Golden Buff and the latest, the Golden Bearded.   In a short time no doubt we shall see the Bantam of all the

Polish varieties. The Standard recognizes two classes of White Crested White Polish Bantams—the Non-Bearded, with white or silver color legs and single comb, which is the original Polish Bantam; the Bearded class, with the V-shaped comb and blue legs. The latter will in time be the recognized Polish Bantam, as it possesses more of the true Polish characteristics. The V-combed birds many think will in time be the only ones allowed to compete in the show room.

Bearded White Crested Polish Bantams.

By "ZIM," for Clough's Bantam Book.

REGARDING the origin of this beautiful breed, and to begin at the beginning, will say that some ten years ago there was an attack made on the Non-Bearded White Crested White Polish Bantams, which, by the way, originated in America. This attack was made by an Englishman whose home is in America, and who had, just previous to the time mentioned, been on a visit to England. He claimed the American production was little short of a cull, and not worthy the name of Polish, in as much as they had *white* instead of *blue* legs, and a *single* instead of a V comb, and asserted that England had produced a White Bearded Polish Bantam that was a miniature Polish. Granted that our non-bearded variety was accepted by the A. P. A. too soon, or rather that they did not form a standard for them that required them to have *blue* legs and V combs, yet they were very nearly or quite as much of a miniature Polish as was the English production, as shortly after his article appeared one of our enterprising Americans imported a pair of these grand birds from England, and the writer of these notes passed judgement on them at one of our large fall fairs, the same season, and was sadly disappointed, as after reading the glowing description of the breed, which at the time he believed, he expected to see a Bantam with all the Polish characteristics, but in reality saw a bird of small size with crest and beard, but with *white legs* and single comb; and as several pairs or trios of our American birds had been sent to England prior to this time have reason to believe they were used in making this *attempt* at Bearded Polish Bantams. After seeing these birds, and remembering the tone of this letter which we had read, which tone savored very much of the superior order of English breeders and brains over American, it was decided that *first* we must have an American standard for Bearded White Polish Bantams which *demanded* they be miniature Bearded White Polish, with all the points and characteristics of the large breed, except size. The Standard description was written and submitted to the A. P. A. and accepted. and there and then *killed* the English attempt on American soil. and nine years ago the *present standard strain* of Bearded White Polish Bantams was started by using birds of the large standard variety on the American non-bearded Bantam, and, by judicious selection and breeding until to-day, we have them as close to perfection as are the large Bearded Polish. with almost perfect V combs, bright blue legs, large globular crests, full beard. and true Polish in shape, and Bantams in size. True, as yet, *prime* specimens are very scarce, and are mostly confined to two flocks, as up until two years ago no eggs were offered for sale. and only an occasional really good bird could be bought. For the last four years the main trouble in breeding them has been the inclination to grow too large, and the failure to reproduce the bright blue leg in all the chicks, but it is safe to say that in two more years neither of these difficulties will bother the breeder of the original flock. The non-bearded variety. when properly handled and well bred, are a beautiful breed, but the bearded birds are in every way their superior. You will note, by this account, that a Standard description of this breed was written by the originator, and accepted by the A. P. A., before there was *a bird in existence* to correspond to its description. and a variety bred to conform to the Standard.

Game Bantams.

IT seems useless to devote space to the Game Bantam in this book, in fact it was the author's first intention not to take them up at all.

Game Bantams are diminutive Games, nothing more or less. The description of the Game varieties apply in every particular to the Bantams, the only difference being the size. If you know what a Game should be then you also know the requirements for the Bantams. Almost all the Game varieties have their Bantams, the most popular of which is the B. B. Red. The Game Bantams are good layers, setters and mothers; they are great flyers and should be kept in a covered run. They can be made as tame as any of the Bantams, and are admired by many as pets.

Every breeder has his hobby, and the hobby of many whom I know is the Game Bantam. In the show room they must be down to the proper weight. The cocks must be dubbed, and any attempt to cutting, trimming or artificially coloring defective feathers will disqualify the specimen.

Scoring Bantams.

MOST breeders believe in the score card; some prefer judging by comparison. There are arguments on both sides. The inexperienced breeder can form an opinion of his own better from the score card than any other method of judging.

In scoring Bantams many times the judge is not as familiar with them as he should be, still a good judge of the large varieties ought to be a judge of the Bantams, as the requirements, while not identical, are nearly so. In judging color there is no distinction; profiles that are correct is what is needed and is what the Standard should contain, and until it does, judging cannot be done satisfactory to exhibitors. In scoring birds judges will vary, sometimes, several points, in fact the same judge will score the same specimen differently at different times. Birds to win first prize are generally required to score 90 points; to win second prize, 88 points; to win third prize, 85 points; and birds not scoring 85 points are not entitled to any prize, even though they are the only specimen in a particular variety on exhibition. In judging fowls in case of a tie score the smallest Bantam should win. In disqualifying any specimen the judge should always give the specimen the benefit of a doubt.

Standard Weights for Bantams.

OTHER THAN GAME.

BREED.	COCK.	COCK'L.	HEN.	PULLET.
Golden Sebright,	26 oz.	22 oz.	22 oz.	20 oz.
Silver Sebright,	26	22	22	20
Rose-Combed White,	26	22	22	20
Rose-Combed Black,	26	22	22	20
Booted White,	26	22	22	20
Buff Cochin,	28	24	24	22
Partridge Cochin,	30	26	26	24
White Cochin,	30	26	26	24
Black Cochin,	30	26	26	24
Japanese (all varieties),	26	22	22	20
W. C. W. Polish,	26	22	22	20

In the show room any bird weighing over the Standard weight will be cut one-half point per ounce for the excess in weight. Should the specimen weigh more than four ounces above the Standard weight, such specimen will be disqualified for over weight (except in Game Bantams which requires six ounces over weight), and will be debarred from competing. In judging Bantams, should two specimens be equal in all point the smallest birds take the prize.

STANDARD WEIGHT FOR GAME BANTAMS.

Cock, 22 oz.;   Cockerel, 20;   Hen, 20;   Pullet, 18.

The above applies to all except Black Breasted Red Malay, which are four ounces more than above.

Disqualifications.

The following defects will generally disqualify a specimen of any variety of Bantams :—shanks other than color of Standard; shanks and outer toes not feathered, in the feathered leg varieties; any feathers on shanks of the clear leg varieties; combs other than specified in the Standard; lopped combs; twisted combs, combs with side sprigs; crooked backs, awry tails; pure white in any part of the plumage of a black variety; pure black in a white variety; birds weighing more than four ounces over Standard weight.

QUESTIONS ANSWERED.

How many varieties of Bantams are there?
There are 14 Standard varieties, and probably as many more that are not yet recognized by the Standard; this does not include the Game Bantams.

What variety is the smallest?
Several are about the same: see Standard weight in this book.

Which variety is considered the best layers?
The Sebrights and Rose-Combed varieties.

Which are the best setters and mothers?
The Cochin varieties surpass all others for this purpose.

Are the Cochin varieties good layers?
A Cochin Bantam is not equal to a Sebright or a Rose-Combed as layers. They lay a rather larger egg and probably three-fourths as many in a year.

Which variety of Cochins are the best?
There is but very little difference. The Buffs are more popular because the color seems to be a taking one.

What varieties are considered as non-setters?
There are no non-setter among the Bantams. The Polish and Japanese do not set as much as many of the other breeds.

Do the eggs from Bantams hatch as well as those of large fowls?
Yes.

Are their chicks difficult to raise?
Not with proper care. They need a little more attention for the first two or three weeks. They should be kept out of the wet grass and fed dry grain such as oatmeal, millet, wheat, cracked corn, etc.

When is the best time to hatch Bantams?
Breeders differ in opinion on this matter; some hatch early, some late, and some early and late. I prefer to hatch in June and July. Very late hatched chicks are smaller, as a rule, but are not naturally so, they are merely stunted. I prefer to breed from small birds that are early hatched rather than small late hatch or stunted birds. If hatched in June or July they will nearly get their growth before very cold weather.

How many can I keep in a coop 10 ft. square?
To allow plenty of room any breed of Bantams should have about 3 or 4 square feet to each fowl or 25 to 33 to a coop 10 ft. square. It would be better however to divide this into two departments of 5 feet each, as small flocks do better than large ones.

What are the diseases of Bantams?
The Bantams are in no way different from other fowls only in their size. They are subject to the same diseases and same treatment. See Dr. Fox's article on diseases.

JUDGE F. B. ZIMMER WAS ASKED THE FOLLOWING QUESTIONS TO WHICH IS ANNEXED HIS REPLIES.

Are Polish Bantams liable to come with feathered legs to any great extent?
No. The original stock *never* bred one with feathers on legs, and I never saw one prior to two years ago, and never since save from stock of one breeder who certainly had introduced foreign blood.

Would they be disqualified if feathered on legs?

According to Standard they could not be. However there should be a clause added, as such a defect is entirely foreign to the breed.

What per cent of Bearded Polish will grow up to be good specimens; *not exhibition*, but good stock and good breeders?

80 per cent, if breeding stock is from reliable source.

What per cent in the Non-Bearded?

90 per cent, on conditions same as above.

Did you ever see *any* or *many* Black Cochin Bantam cocks or cockerels that were *black* to the skin, and free from any white in under color?

Have never seen but two *cocks*, that had moulted, that were clean, clear black. Have seen several cockerels, but 80 per cent will have neck under color light.

If you have, what do you consider such a bird worth?

Cock, $25 to $50; cockerel, $10 to $20.

Do you prefer white or yellow legs in White Cochin Bantams?

Yellow.

Which do you prefer in Rose-combed White?

White with the pinkish or flesh colored tinge.

In all the Cochin varieties, except Buff, the Standard disqualifies for lopped combs. In the Buff it disqualifies for twisted combs. Now would you disqualify a Buff for lopped comb?

No. However, I believe the Standard should be the same in all varieties.

Would you disqualify the others for a twisted comb?

No. The "twisted" and "lop" I consider should be the same in disqualifications of all the Cochin Bantams. Either sort of comb should disqualify any of the family of Cochin Bantams.

The Standard disqualifies all the large Cochin varieties for bare middle toe, but not the Bantams. In your opinion will they do so in the next edition of the Standard?

Hardly think it will come to this so soon. However it would materially help the leg and toe feathering on them if it did, and I should favor adding it to the Bantam disqualifications.

Does the Black Tailed Japanese breed fairly true?

Only fairly; they require great skill to produce prime specimens.

Of all varieties of Bantams which lead in numbers at our shows?

Buff Cochin, Sebrights and Black Red Games.

Which varieties are seldom seen?

Rumples, good White Rose Combed are scarce, Partridge Cochin, Light and Dark Brahmas.

The Standard gives color of ear lobes in the Sebright as "immaterial." What color is preferable?

Breeders pay but little attention to this matter. An answer from me would be only my private opinion. *I* prefer *red*. Why? Simply because I *never* saw a bird with white lobes that was worth over $2. Always poorly laced, and *white lobes* is simply *nothing* as compared to a well laced bird with good shape.

Is red and white (mixed) regarded as inferior to a solid color in the show room?

No. Color is immaterial, only see that the lobes are smooth, free from wrinkles, of proper size, shape and texture.

Are the Polish Bantams more delicate (the chicks) than other varieties?

No more than several of the other varieties. However they are easier prey for hawks; cannot keep themselves as free from lice as those breeds devoid of crests, and therefore may require more care from the breeder.

The Diseases of Bantams.

THEIR PREVENTION AND TREATMENT.

By DR. FOX; FOR CLOUGH'S BANTAM BOOK.

BANTAMS are subject to the same diseases that their larger brothers and sisters are, and it is often said that they are more delicate and more liable to disease than the large varieties. If this is true, it is not because of their small size or any natural weakness, so much as it is because of the methods used in rearing them. For many breeders keep breeding in and in for generations, to keep down the size, then, too, they hatch late in the season and feed very light, in fact keep the growing chicks on the verge of starvation all the time. The result of such methods are, of course, frail, delicate fowl, which are liable to all kinds of diseases. Any of the large breeds treated in the same way become just as delicate.

This way of keeping Bantams small is not only wrong, but is entirely unnecessary, for there is no variety of Bantam that can not be raised in a perfectly healthy and natural manner, and still kept down to weight required by the Standard, and Standard weight is small enough. No doubt single specimens can be and, in fact, are reduced to one-half Standard size; but to what end? They lack vigor and rarely have the perfection of shape and plumage which we want. If breeders would give less attention to reducing weight and more to producing Standard shape and fine plumage, we should see Bantams shown in better condition, and should be less troubled with disease.

The object of this article is not to give an exhaustive treatise on the subject of diseases in Bantams, but to show the novice how he can have a flock of strong, healthy Bantams and how he can keep them so.

First and foremost, he must get as vigorous stock as possible to start with. This will depend largely on the breeder he buys of. There are many reliable men who are now, and have been for years breeding Bantams carefully and intelligently, and who have strong stock. Almost any one who is thinking of starting into the breeding of Bantams must have some friend who is posted in such matters and who will gladly recommend such stock as he knows to be good.

When buying a pair, trio or pen of Bantams to breed, be sure the cock is not related to his wives nearer than second cousin. A breeder who raises many of one variety will generally be able to supply birds not too nearly related. Perhaps the surest way, however, would be to buy males of one man and females of another. Bear this question of in-breeding in mind every year when mating your pens, if you want good, vigorous chickens.

It is true that a certain amount of in-breeding is necessary to attain perfection in fancy points and is practiced by almost all experienced breeders, but it is a dangerous practice for an amateur and, at the best, is sacrificing vigor and strength to show room qualities. Leave the in-breeding to men who know just how far to go, and who can see the first danger signal, and let your matings be of birds not related to each other. This talk may seem foreign to the subject but is, in reality, of the first importance, for there is no one thing that tends to produce stunted, unhealthy specimens so much as the reckless mating of relatives year after year.

Having got healthy stock, how is the amateur to keep it so? In the first place, give the little birds good quarters. Now, the house need not be expensive, handsome or very large, but must be light and dry or there will be no success. It must also be clean, and *always* clean. Don't think it is enough if the hen-house is hoed out in the fall and spring. It must be cleaned every single day in the year. It is really easier to do this every day than it is once or twice a month, and if the house is conveniently arranged it will take less actual time. No man lets his horses or his cows stand in their own filth day after day; then why, pray, should the poor hen be obliged to roost over piles of stinking manure?

Have a good wide dropping board under the roosts, and keep it covered with earth or fine sand; with a fine sieve in one hand and a small coal shovel in the other, the droppings can be cleaned up in less time than it takes to write about it. In summer, when the fowls are in the yard all day, the floor can be kept sprinkled with sand and will require very little attention; a sweeping and and sifting every week or two will do very well. This is supposing the floor is a hard one, either cement or boards, as it should be. In the winter the floor must be covered with scratching material, as dry leaves, sweepings from the hay mow or broken straw, and this should be renewed every few days, as it soon gets foul. A better way is to have a scratching shed adjoining the roosting room which can be kept just as in summer.

In arranging a house for the pets the question of ventilation will probably come up. There has been a vast deal written on this subject in all the poultry papers, and there is a great difference of opinion about it. What ever may be the best plan for large varieties the following has certainly proved the best for Bantams. Build a house as near air-tight as boards and roofing paper will make it. Have a door and a small window in the south side, with the roosts on the north side, so as not to be in a draught when the window is open. On summer nights have the window and door wide open. As cool nights come on begin to close up until, in severe winter weather, everything is shut tight at sunset. Don't bother about ventilation; there will be more than enough air get in through the walls and cracks and if the house is kept clean the Bantams will not suffer. Of course if the house is filled with the stench of accumulated droppings, shutting it up tight makes a bad matter worse. When the thermometer is up above freezing, leave the window open a little at night, but when below freezing shut everything. In the middle of the day, if the sun shines, open ths window for a short time, even on very cold days.

It is now pretty generally understood that poultry must have exercise, if eggs are wanted in the winter, but this is a point seldom thought of in regard to Bantams. While we probably do not expect them to lay much in the winter, and in fact prefer to have the eggs come at hatching time, still, if we wish to keep the little fellows in the best condition we must give them exercise. This will prevent feather pulling, egg eating and similar vices.

The only way to provide this exercise, in winter, is to give them plenty of litter and make them scratch for every kernel of grain they have. In summer give them a grass run and they will keep on the move all the time.

Some care must be used in feeding Bantams, as well as large fowl, to get good results. In the summer if they have a grass run the problem is very simple; in the morning give oats, and once or twice a week mix a little cracked corn with the oats. This morning feed should be rather scant, so that they will be hungry enough to keep after the insects and worms to be found in the grass, all day. At night give a full feed of wheat, that is, all they will clean up in ten or fifteen minutes. Keep pure, clean water and good sharp grit before them all the time; do not let the water stand in the sun. In the winter there is more difficulty, as the green food and animal food have to be provided artificially. For green food give them cabbages, turnips, potatoes, onions, apples, etc., a little every day. The animal food may be provided by giving green cut bone three times a week, or if this can not be had, a mash containing animal or blood meal may be given. This mash should be composed of two parts animal meal, one part corn meal, two parts shorts and two parts ground oats; the whole to be stirred into boiling mashed potatoes or boiling cut clover, or better still, boiling table scraps. This must be mixed hard, not a bit sloppy, and may be fed warm or cold. The grain must be all scattered in litter, and the birds should be kept hungry enough so that they will scratch lively. Overfeeding is to be carefully avoided.

A good bill of fare in the winter is a mash for breakfast four mornings a week, cracked corn and oats, half and half, the other three mornings; at noon some green food, and at night a good, full feed of wheat. Keep grit and oyster shells before them all the time and, of course, plenty of water. If you have a bone cutter you need not bother with the mash as the cut bone can be fed three or four mornings a week, in the place of the mash. A bone cutter will soon pay for itself in saving of grain, and cut bone is the best of food for poultry.

Be sure the little hens have a chance to dust, as it is their only way of keeping clean and free from lice. Just watch them as they take their dust bath and see their evident enjoyment, and I think you will be sure to supply them with it every day in the year. In the summer there is usually a dry hole in some corner of the yard that answers the purpose, see that the earth is kept soft and not allowed to cake up. In the winter a dust box must be provided in a sunny part of the house or shed. This box should be big enough to accommodate two or three at one time and should be filled with dry earth or road dust, and a little sulphur mixed with it.

The feeding and care of Bantam chicks is not very different from that of other chicks. Still this article will be incomplete if it does not tell you how to raise healthy chickens. Leave the hen and her chickens on the nest until all have hatched and they are all dry. This will usually be about twenty-four to thirty-six hours after the first one hatches, and about as long as the hen will stay on. She must be taken off before she tries to get the chicks onto the floor, as some are pretty sure to get hurt if she attempts it. Put the hen and her brood into a clean, dry coop (an exhibition coop is just the thing), have the floor thickly covered with sharp, fine sand, and then throw in a little finely broken egg shell. Have a suitable vessel of water in the coop, and don't forget to give the hen a feed of corn, otherwise she will be restless and uneasy.

For the first week feed dry bread crumbs, oat grits or fine head oat meal, as it is sometimes called, and cracked wheat. Feed everything dry and throw it onto the sand so they will get enough grit. Feed every three hours, all they will clean up. Don't keep food before them all the time. Corn bread, graham bread or any kind of breakfast gems are good to crumble up. They should be dry enough to crumble fine, but not musty. If you cannot get enough bread, etc., make a cake as follows: two parts corn meal, two parts ground oats, two parts shorts and one part animal meal, stir up with water and a little baking powder; bake thoroughly and feed in crumbs.

At the end of the first week set the coop on the grass and let the chickens run all they will, but do not let the hen out yet.

Feed the same as the first week with the addition of finely minced meat, well cooked, once a day. Along in the third week, if the chicks seem pretty strong, let the hen out about an hour before sunset and gradually lengthen the periods of freedom until, when the chicks are a month old, she is out all day. The point is not to let her tire the little fellows all out by keeping them on the move all day. About the end of the third week you can begin to feed whole wheat and cracked corn, and in the place of the meat a little cut bone.

Bantams feather much quicker than the large breeds and must be fed very generously until fully feathered out, and meat and cut bone are important. Most varieties are pretty well feathered when seven or eight weeks old and after that will not need to be fed so high. If they have a good grass run two, or at most, three feeds a day is enough, and they should consist of cracked corn, wheat and oats. When they begin to get their second feathers, along in the fourth or fifth month, look out for them and feed meat and cut bone, as this is a critical time and you will lose some if you stint their food the least bit. Many will advise you not to feed corn either to chicks or fowl, for fear of making them too large. This fear is without foundation, and corn should form a part of the ration for both chicks and fowl, as they can not be kept in the pink of condition without it.

If you will follow carefully the directions already given you will have little need for the remainder of this article. Still, even with the best possible care, there will be an occasional invalid in the flock, and I will try and tell you what to do in such a case.

It is of the utmost importance to be able to detect the presence of disease at once, when it appears, not only in order to get the best results of treatment in the individual case, but prevent the spread of contagious diseases. This one point is the worst stumbling block that the inexperienced breeder will meet with, as no amount of book knowledge can teach him to be quick to see and understand the first symptoms. Nothing but experience and, sometimes, bitter experience will give him this faculty. He must learn first to know his fowl in health, their movements, their appetites, the appearance of their combs, their carriage, etc. Then, when there is anything which varies from this healthy condition, he must learn to know what it means. The following description of the most common diseases may be of assistance in this respect. More attention will be given to the prevention than to the treatment, for one ounce of prevention is worth five pounds of the latter in the diseases of Bantams.

COLD.

A common cold is probably the most prevalent disease the human family is subject to, and the same is true of Bantams. The first symptom is sneezing, then a discharge of clear, watery fluid from the nostrils and eyes; later, a slight loss of appetite and a general dumpishness.

In itself a cold is of little consequence, but, as it is often the forerunner of roup, it must not be neglected. Cold is generally caused by draughts blowing across the roosts at night, or by filthy quarters. It may also be caused by dampness in the house or runs, or too much exposure to bad weather. Bantams can be allowed in their yards in very cold weather if the ground is free from snow and mud, but they are much better off in the house if there is mud or snow on the ground or if it is stormy. In this respect they certainly require more care than the large varieties.

The prevention of colds lies in keeping the flock in clean, tight, dry quarters.

The treatment is very simple. If only one or two are affected, remove them from the rest and place in a coop where they will be warm and free from draughts. Get some camphorated oil, at any drug store, and, with a small glass syringe, inject it into the nostrils twice a day. This will generally effect a cure within a few days. If many of the flock are afflicted in this way it will be impracticable to treat separately, and the first thing to do is to find and remove the cause of the illness. Having done this, keep a small piece of gum camphor in the drinking water and watch carefully for further symptoms. Do not allow the nostrils to become plugged by a crust, as it often will, because the discharge will be held back and act as a poison.

After the nostril has been obstructed a day or two the head will begin to swell and before we know it we have a case of roup

to deal with. The injection of camphorated oil as already direct-
ed will usually keep the nostrils free and open.

ROUP.

This is a contagious disease, and generally begins as a simple
cold. It is often fatal, and is much to be dreaded as it will sometimes go through the whole flock before the owner is aware that
there is any serious trouble. It is difficult to say just when a cold
turns into roup, but when the discharge from the nostrils and eyes
becomes thick and sticky, and of an offensive odor, you may be
sure you have a case of roup. The next symptom is swelling of
the head and eyes; frequently the eyelids will be stuck together,
and if washed apart a large amount of fetid matter will escape.
As these symptoms increase the bird is growing sicker all the
time, more dumpish and has little or no appetite.

Roup may be prevented by good care and by prompt treatment of every cold, but above all by care in introducing new birds
into the flock. Whenever you buy a new hen keep her in quarantine at least two weeks, until you are sure she is in perfect
health, before exposing your stock to the danger of contagion.
Bantams of a strong, vigorous constitution, properly housed and
fed, will never have roup, unless they catch it from some diseased
fowl carelessly introduced into their house.

Probably the most common way for the disease to be trans-
mitted, from one to another, is through the drinking water. Be
careful to thoroughly clean and scald any drinking vessel that
has been used by any sick Bantam, before using it again. It is
doubtful whether the disease can be carried in the air, but give
the well birds the benefit of the doubt and confine diseased ones in
separate houses or rooms. It is unwise to keep an invalid in a
room with a fire, unless you are prepared to keep him there until
warm weather, for it will never be safe to return the convalescent
patient to the unheated house after he has had the luxury of a fire.

The treatment of roup is, in the main, very unsatisfactory,
although, if begun soon enough, it may save a valuable specimen.

Keep the nostrils, eyes and throat as clean as possible. Get
a bottle of listerine at any drug store, and put a tablespoonful into
a glass of warm water. Inject this into the nostrils, swab the

throat and wash the head and eyes with it two or three times a day. Give one drop of tincture of aconite three times a day for the first four or five days. Feed with soft cooked food and milk.

If this treatment makes no improvement in the patient in a week, kill him and burn his carcass. This is the kindest and best advice that can be given, for, although he may recover after weeks of dosing and pampering, he will still be a weak bird and the slightest exposure will start a discharge from the nostrils, which may contain the germs of roup and be sufficient to cause the disease in the flock to which he belongs.

A Bantam that has once had a genuine severe attack of roup is never fit to breed from, as his offspring will be sickly, puny chicks nine times out of ten. If you are unwilling to take this advice, as you probably will be until you have tried to cure roup yourself, the next best thing to do is to continue to keep head and nostrils as clean as possible. Stop the aconite and give one grain of sulphate of quinine three times a day, and all the milk and whiskey you can pour down every three or four hours. By this time your pet will not eat and his strength must be kept up by forcing the whiskey and milk. Should your efforts prove successful and the bird begins to mend, leave off the whiskey and quinine very gradually and put enough tincture of chloride of iron into the drinking water to give it a decided brown color; feed good cooked food and a little meat once a day.

CANKER OR DIPHTHERITIC ROUP.

This is a frequent accompaniment of ordinary roup, and is probably a different manifestation of the same disease. It is highly contagious to other fowl and possibly to man. Cases are reported where children have probably contracted diphtheria from fowl sick with canker, and also where poultry that have had access to discharges from diphtheria patients have sickness with canker. The one distinguishing symptom of canker is the appearance in the mouth or throat of a white or yellowish white cheesy membrane. This may appear during the course of ordinary roup, or may came on suddenly in an apparently healthy fowl. At the first onset one or more white spots, about the size of a pin head, may be seen either on the roof of the mouth or under the tongue or, quite often, around the opening to the wind pipe. These spots grow very rapidly until, often times, the whole mouth is filled with a membrane that is usually glistening white, sometimes yellowish. When torn off it leaves a bleeding surface beneath. It is of very offensive odor. If this membrane extends into the wind pipe the patient will soon die of suffocation. This is a disease that cannot be mistaken, as the appearance of the membrane is very characteristic.

The remarks on the cause and prevention of roup apply equally to canker and need not be repeated.

The general treatment is also the same, but the local treatment is different. Instead of washing out nostrils and mouth attempts must be made to remove the membrane. This is often done by scraping with a piece of pine wood whittled to a convenient shape. After removing all that can be removed, without excessive bleeding, the parts should be powdered over with alum. A better way is to apply peroxide of hydrogen in full strength directly to the membrane, which will soon be eaten away with much less bleeding than in the other proceeding. After using the peroxide a few minutes, apply tincture of the chloride of iron in full strength. The mouth can be pretty well cleaned by either method, but the membrane soon returns and the process must be repeated often. When the membrane is in the wind pipe it has to be left to nature, and almost always proves fatal.

CHOLERA.

At the present day this is an extremely rare disease in the United States. It is the most contagious of the diseases of poultry, generally killing the whole flock when it once gets a foothold. It is always caused by contact with a previous case, never originating in a yard without such contact or exposure.

The symptoms are excessive diarrhœa, first of a black substance as thick as tar, later by a thin watery fluid which smells putrid. There is very rapid emaciation and prostration, death frequently occuring within thirty-six hours after the commencement of the disease. There is no treatment; kill and cremate.

DIARRHŒA.

This is quite frequent and is sometimes mistaken for cholera, but cholera is so very rapid that this mistake ought not to be made.

Diarrhœa is usually caused by improper food, impure water, by sudden changes in temperature or exposure to cold and wet. Individual mild cases require no treatment as they will soon recover. In severe cases remove the patient to a coop, keep without food for twenty-four hours, keep lime water before it instead of clear water. After twenty-four hours give a little bread soaked in boiled milk. Let this be the only food until diarrhœa ceases. When there are a number of cases in the flock, be sure there is something wrong in food or drink. Search carefully for this cause and remove it.

CROP BOUND.

This is quite common in Bantams and if not properly treated is very apt to prove fatal. The first symptom is a constant effort to swallow. The neck is stretched out, the mouth opened and the hen acts the way you often see a little chick act, when trying to get down a worm one size larger than his gullet.

The patient acts dumpish and stands in a peculiar position, with the breast bone pitched forward and down. He is hungry and will keep eating until his crop is filled full and as hard as a stone. If you suspect that you have a case of crop bound place the subject where he cannot eat for twenty-four hours and then feel his crop; if it is as hard, or harder, then when he was shut up your suspicions are confirmed.

This trouble is caused by a plugging up of the outlet of the crop with some particle of food, such as a long ribbon-like piece of hay or grass. It may be caused by over eating, as when fowl get access to the grain bin and then drink a lot of water. The cause in this case is, probably, not so much obstruction of the outlet as it is a paralysis of the muscles of the crop from over distension. This is a rather unusual form of crop bound, and is merely mentioned to point this moral; when you know your Bantams have enormously overeaten, deprive them of water until their crops are, at least, half empty. There is no way to prevent the other or obstructive form.

The treatment is the same in either case; empty the crop. This can sometimes be done by pouring castor oil down the throat and working the mass in the crop around with the fingers. Try this about three times, two or three hours apart. If by that time the mass is not softened it is time to resort to surgery. Remove the feathers from a space the size of a silver dollar, directly over the crop. With a clean sharp knife make a cut one and one-half inches long through the skin; pull the wound along about half an inch, and with a second cut go directly through into the crop. With a spoon handle scoop out the contents of the crop thoroughly. Either see or feel the outlet of the crop, so as to remove any obstruction there may be there. Wash the inside of the crop and the wound with warm water, to which a little salt has been added. With a needle full of white silk sew up the crop and then the skin.

Give no food or drink for thirty-six hours, then give a little bread soaked in milk. Feed carefully for a week; by that time the little fellow will be all right, that is, supposing the relief to have been given soon enough. For, if the mass in the crop has fermented badly, as it will in three or four days, it will have excited so much inflammation that the operation does no good. Do not delay in a case of crop bound as twenty-four hours frequently makes the difference between saving and losing a valuable bird.

LEG WEAKNESS.

This is most common in growing chickens and is shown by inability to stand up. The chicken appears hungry, and all right in every way, except that it tries to get around on its hock joints instead of its feet. This occurs either while the first feathers or the second are growing. It is due to defective nutrition and is analogous to what we frequently term in children "growing too fast for their strength." The remedy is to change the diet, giving more meat and cut bone, something to make more muscle. Take care that the other chicks don't prevent the weak one from getting any food at all. With a little care these cases recover in a few days.

In the full grown Bantams a similar condition is often seen, although not so often as in the heavy breeds, and is more apt to be due to rheumatism or cramp, the result of dampness or exposure. The remedy in these cases is to place the patient in a dry coop and feed well, at the same time rubbing the legs well with any good liniment.

SCALY LEGS.

This is a most disgusting affection and its presence in a flock is a sure sign of laziness or indifference on the part of the owner. It is caused by a parasite, and is, therefore, a contagious disease. When it first appears the shanks and toes become covered with a dry scaly substance which increases quite fast until it forms crusts so thick as to obscure entirely the original shape and color of the legs. It is most common among the feathered legged varieties, and spreads much faster in damp, filthy quarters than in clean and dry ones.

The treatment is very simple, but is also very effective. Apply thoroughly, with the fingers, some carbolized vaseline to every part of the shanks and toes. Repeat every two days until the legs are clean. Each time it will be found that considerable scale may be rubbed off with the fingers, and it is advisable to remove all that will come off without causing bleeding. In mild cases three applications is enough to effect a cure. In severe ones it may take six or seven, and, in such cases, it is well to repeat twice a month for three or four months after the case is apparently cured, as it otherwise is very liable to return.

LICE.

If you have had no experience with poultry you will probably smile when you see lice classed among the diseases, but after one or two broods of chickens have succumbed to their ravages, and the grown fowl all look as if they were in the last stages of consumption, you will admit that the little vermin are worthy of the first place in the list of diseases.

There are several varieties of lice which infest the hen house. There is the common white or grey louse which is the largest and stick to the fowl day and night. The same variety is found on young chicks and is commonly called the head louse because oftenest found on the head and fastened to the skin like a leach. Then there is the red louse, or red mite, which works only at night. During the day he will be found under or on the roosting pole, or on the sides of the house. He is bright red, round and rather smaller than the head of a pin. Frequently these mites will congregate on a part of the wall so thick that one would think the wall was covered with fresh blood.

There is also a brown louse, larger than the red and not so large as the white. The habits of this are similar to both the others, that is to say, many will leave the fowl in the day time and be found in the house, but some of the more greedy will keep at work day and night. This is the kind that bothers the sitting hen the most, sometimes she is compelled to leave her eggs, and, in such instances, one looking into the nest will see no eggs there as they will be completely covered with a mass of the dirty brown lice.

The symptoms produced by lice are unmistakable, where one has once become acquainted with them. In fowl there is ruffled plumage, white comb, great uneasiness and emaciation. In chicks there is weakness and drooping, sometimes diarrhœa and a peculiar, characteristic look about the head, as if the beak had been pulled on and the head elongated. The proof that the symptoms are caused by lice is to see the enemy.

In this connecton a very good answer appeared in the notes and queries of a recent poultry paper. The question was like this: "What is the matter with my chickens, they have such and such symptoms?" Answer, "Look for lice, and if you find them remove by doing thus and so. If you don't find any do just the same, for they are there, only you don't know how to look for them."

In looking for lice on fowl, look close to the skin around the vent and under wings; on chicks, examine head and under wings; in the house, look on under side of roost and into all the cracks and crevices.

The prevention and treatment are identical. Keep dropping board clean in hot weather, sprinkle slaked lime over it occasionally.	Have the roosts and dropping board arranged so that they can easily be removed.	Take them out in the yard twice a month, in summer, and paint them all over with kerosene, at the same time paint walls and cracks near where roosts belong.	That same night go into the house and sprinkle a little Lambert's Death to Lice over the back of each hen.	Clean out the nest boxes and paint inside and out with kerosene.	Refill with clean nesting material and sprinkle a little Lambert's Death to Lice in it.	Never set a hen without dusting both her and the nest thoroughly with the same powder, and repeat at least three times while she is sitting.	When the chicks hatch, welcome them with a good dose of Lambert's, and repeat, at least, once a week for the first two months of their lives.	There are probably other insecticides as good as Lambert's Death to Lice, but I have never seen them, and, as I know that that will do the work, I don't hesitate to recommend it.

If the chickens are badly infested with head lice, the quickest way to relieve them is to apply a very little vaseline to the top of their heads and under their wings.	After one application of this the free use of Death to Lice will keep them away.	Do not forget to keep the chicken coops clean, as filth is the very best breeding place for lice.

You cannot breed Bantam chicks and lice in the same place and at the same time.	If you doubt this give a lousy hen a nice brood of the little chicks and see the result.	After one practical lesson of this kind the most skeptical will be willing to go to a great deal of trouble to keep his chicks free from vermin.

GAPES.

This is an affection seen only in young chicks from the third week to about the third month.	It is, fortunately, not common in moderate climates, although said to be quite prevalent in the South.

Gapes is caused by the presence in the wind pipe of one or more thread-like worms.	These little worms attach themselves to the lining membrane of the wind pipe and cause it to swell, so that it fills the whole calibre of the pipe, and the chick dies from suffocation. The principle symptom is gaping. The chick stretches his neck and opens his mouth to its fullest extent. He does this repeatedly and soon refuses to eat, becomes dumpish, and, if not relieved, dies. The only preventative is absolute cleanliness about the coops and yards.

The treatment of gapes is not very satisfactory. It consists in removing the worm from the wind pipe. This can be accomplished by means of an instrument known as the gape worm extractor. The operation requires some skill and more patience. When a large number have to be treated the treatment is wholesale, so to speak, and the usual method is to smoke the rooms out. The chicks are shut in a tight box which is then filled with the fumes of burning sulphur or carbolic acid, or with finely powdered slaked lime. The trouble with this method is that the worms will stand about as much of it as the chicks will, and you will be very lucky if you can stop at just the right moment, that is, when the worms are killed and before the chicks are. Chickens that have had gapes are feeble and debilitated for a long time, and perhaps you will be more lucky, on the whole, if your smoke kills both chicks and worms.

Better direct your energies to stopping the spread of gapes than to doctoring those already affected. Take all the sick and place them in a clean, dry coop with sand and air-slaked lime on the floor. Take the rest of the brood and all the chicks that have had access to the same yard, put them into quarters by themselves and watch very sharply, so as to remove each one to the hospital coop as soon as it shows a symptom. Be sure that any chicks that have not been exposed to danger are kept away from the infected yard, from the quarantined chicks, and, of course, from the sick ones, until the disease is thoroughly stamped out.

The infected coops and yards must be disinfected. A good way to do this is as follows: burn all old coops that are not of much value; mix a hogshead of corrosive sublimate of strength 1 to 2000; heat to boiling point enough of this solution to saturate every part of the coops. Sprinkle the rest of the solution over the ground. When the coops are dry give a good coat of whitewash. Sprinkle air-slaked lime over the ground until no earth can be seen. Leave alone for two weeks and then spade and sow down to grass. Put no chicks into this yard for two years. Fowl may be kept in it after the grass is grown, if necessary, but no chicks.

PIP.

This is a disease of young chickens and is practically a cold. It occurs oftenest in chicks whose parents have had roup, or have been in-bred too much. It is sometimes caused by damp and filthy coops.

Treatment; give dry, clean quarters, and wash mouth and nostrils with a weak solution of chlorate of potash.

CHICKEN POX.

This is a highly contagious disease which affects both old and young. It is rather rare. It is characterized by black, hard warts or growths on the head and face.

The only treatment is to quarantine and keep the warts greased well with carbolized vaseline. Fowl will generally recover and be as good as ever; while chicks almost always succumb within a week or two after they are taken.

GOING LIGHT.

This is not a very definite term, and the condition to which it is applied is also called consumption, scrofula, congestion of the liver and inflammation. It occurs occasionally in flocks that have the best of care, so it seems there is no sure way to prevent it.

It is undoubtedly a disease of the digestive organs, possibly the liver. Autopsies often show a liver rather too large, but no other abnormal condition visable to the naked eye.

The symptoms are great emaciation, extreme pallor of the face and comb, ruffling of feathers and general dumpishness. During the first of it the appetite is fairly good, but later disappears entirely.

When the disease attacks a chicken that is getting its second feathers, as it often does, it is, as a rule, fatal. To be of any avail treatment must be begun very early. Give sulphate of strychnine, 1-200 grain, three times a day, and color the drinking water with tincture of chloride of iron. Feed meat, green food and some cooked food, as bread or mash.

When the patient is a grown fowl the treatment is somewhat different. Shut in a coop with clean sand on the floor, give calomel, 1-10 grain, every two hours for five times, and no food of any kind, but plenty of water. The next morning after these five doses the droppings should be found, in the sand, abundant and rather loose, if they are not give a level teaspoonful of epsom salts. After the bird has been well physicked in this way, begin to feed soft food rather sparingly until your patient seems really hungry. Give the strychnine and iron, as in the previous case. As soon as the appetite returns put her back in the run where she can get more exercise and variety of food. Watch her carefully and if she grows worse again repeat former treatment of calomel. It is often necessary to do this three or four times before thorough recovery takes place.

Now, in conclusion, just a word. Remember that you will be well repaid for all the time and pains which you care to spend in giving your Bantams all proper care to keep them in good health. On the contrary, in nursing sick Bantams your time will frequently be thrown away. The moral of this is: do your best to prevent disease, and when it does appear, as it sometimes will in spite of your best endeavors, don't be afraid to use the hatchet.

DR. WILLIAM Y. FOX,.

BREEDER OF

WHITE, BLACK AND BUFF   COCHIN BANTAMS,

TAUNTON, MASS.

I have bred these little pets for a number of years, but did not exhibit until December, 1895, when my birds were winners at Providence, R. I., scoring from $91\frac{1}{2}$ to $95\frac{1}{2}$; and at the great Boston Show, January, 1896, I won eight ribbons with eight entries.

I have been especially careful to keep up the vigor of my stock, and at the same time have not sacrificed size, as my Bantams are well down to standard weight.

I am now breeding them in large numbers, and am prepared to fill orders for stock at reasonable prices.

No eggs for sale.

CUT THIS OUT.

Poultry Topics,
        Warsaw, Mo.

        Please send to my address below, several back numbers of POULTRY TOPICS for my inspection (without charge).

Name

P. O.   State,

Mail to

POULTRY TOPICS, WARSAW, MO.

For the Poultryman.

THE CROWN BONE CUTTER

BEST IN THE WORLD.

PRICE,

Like illustration $6.50, with stand $8.50

Weight, without stand 50 lbs., with stand 80 lbs.

THE GEM CLOVER CUTTER

Cuts fine fast and easy

Has steel knives and adjustable cutter bar.

The $5 Hand Bone.

Shell, Corn and Grit Mill.

Send for circulars and testimonials.

WILSON BROS.,

EASTON, PA

TESTIMONIALS.

Woodlawn, New York City, Oct. 8. 1896.

Messrs. Wilson Bros., Easton, Pa., Gentlemen: I am very much pleased with the "Crown" Bone Cutter. I found it very satisfactory. Yours very truly, JOS. T. HILBERT.

Caroline, N. Y., October 5, 1896.

Messrs. Wilson Bros., Easton, Pa., Dear Sirs: The "Crown" Bone Cutter I purchased of you last winter has proved a complete success. Is the easiest running and does the best work for its size of any I have ever seen. Through its good work my father-in-law, Theo. Steenberg, of South Danby, N. Y., also purchased one. Yours very truly, W. G. KRUM.

Reading, Pa., October 5. 1896.

Messrs. Wilson Bros., Easton, Pa., Gentlemen: Having in use one of your Shell Crushers and Crown Green Bone Cutters, I find them indispensible for breeders of poultry. Your Crown Bone Cutter is the cheapest and lightest running machine I have found and can recommend it to all fanciers in need of such a machine. Through my machine I have caused the sale of five others. Yours, F. A. SCHOFER, Breeder of High Class Fancy Pigeons.

GRAHAM'S

AMERICA'S BEST

Buff Cochin Bantams.

ALWAYS WINNERS

At New York, Boston, Washington, Hagerstown,
Trenton. Mt. Holly, Gloucester, Hackensack, N.
J., West Chester and Reading, Pa., Meridan
and Stafford Springs, Conn.

FINE EXHIBITION STOCK FOR SALE.
NO EGGS.

SEND 4¢ FOR CATALOGUE.

LOUIS P. GRAHAM,

1740 So. 16th St.,   *PHILADELPHIA, PA.*

THE POULTRY GRAPHIC

A 20-PAGE MONTHLY JOURNAL DEVOTED TO THE BREEDING AND REARING OF POULTRY, PIGEONS AND PET STOCK.

------ ONLY 25 CENTS A YEAR. ------

SAMPLE COPY FREE.

It's a Winner. It is progressive and up to date. It is well edited by a practical poultryman. It is nicely printed and illustrated. Its contributors are of the very best. Each number contains information of inestimable value to everyone who is interested in Poultry, Pigeons or Pet Stock.

It is absolutely the best advertising medium (cost considered) for poultrymen in the United States. It has a faculty of reaching the buyers. Its rates are way down. It brings results. It makes business. It reaches every state and territory in the Union. It's all right. It's a Winner.

THE POULTRY GRAPHIC,

J. F. SCHUREMAN, JR.,
Editor and Publisher.

GENESEO, ILL.

Do you need a Bone Cutter?

SENT ON TRIAL!

If so try the

"Dandy"

This Cut shows our $7.00 Machine.

Same thing without balance wheel, $5.

We make 5 sizes, $5, $7, $10, $12 and $18.

What the people who have used the "Dandy" say about it
is what tells the story.

Send for Illustrated Catalogue and Testimonials.

STRATTON & OSBORNE, Erie, Pa.

Mention this book.

TESTIMONIALS.

Meadville, Pa., Sept. 6th, 1895.
Messrs. Stratton & Osborne, Erie, Pa.: I received the Dandy Bone Cutter all O. K. It does its work all right and I am pleased with it.   Yours truly,   T. H. APPLE.

Ashley, O., Jan. 24. 1896.
Messrs. Stratton & Osborne, Erie, Pa.: The Bone Cutter ordered of you came in good shape, and after testing it am well pleased.   Yours truly,   E. JOSEPH COLE.

Bridgewater, Mass., Oct. 13th, 1895.
Messrs. Stratton & Osborne, Erie, Pa., Dear Sirs: Received the Dandy Bone Cutter in good shape, and am well pleased with it. It does better work than any bone cutter I have ever seen.   Yours truly,   W. N. COOKE.

COTTAGE ST. POULTRY YARDS, R. R. Hamilton, Prop.
Peabody, Mass., Dec. 16th, 1895.
Messrs. Stratton & Osborne, Erie, Pa., Gents: The Bone Cutter arrived all right. I have seen a good many bone cutters, but never saw one to beat the "Dandy." I have given it another name, I call it a little engine. It cuts very fast and fine. I can sit on a chair and turn it. I have it up stairs in my poultry building, and am proud to say I own a "Dandy" Bone Cutter. When it arrived, I set it up and started it. My wife wanted to try it, so I filled the box and she turned it awhile; "Well," she said, "you have a bone cutter at last." The ——bone cutter I had before was a man-killer. How much does a set of knives cost?
Yours truly,   ROBERT R. HAMILTON.

AMERICAN POULTRY JOURNAL,

**325 Dearborn St.,
CHICAGO, ILL. . . .**

A Journal

that is over a quarter of a century old and is still growing must possess intrinsic merit of its own, and its field must be a valuable one. Such is the American Poultry Journal.

What Part are you going to get?

YOU

can't get all the business in the country but the choicest portions are yours if you hustle. Let the people know where you are; what kind of stock you keep; the prices, etc. Put these advertisements where poultrymen must see them—in the

American Poultry Journal,

**325 Dearborn St.,
Chicago, Ill.**

50 cts. a year.

'CARBOZIN'

The Greatest Discovery of Modern Times for the Restoration and Preservation of Health in Feathered Animals of all Kinds.

"CARBOZIN" is a vigorous disinfectant, alike efficient for man, for beast and for bird.

For the present we address ourselves to those interested in the raising of poultry, and we offer "CARBOZIN" as the cheapest and most positive vitalizer and invigorator known to the trade. One dollar spent in "CARBOZIN" is worth ten, easily, in the increased health and vigor, and consequent productiveness of even a small lot of poultry.

"CARBOZIN" is easily applied. Sprinkle the nests lightly, and paint the roosts, sides and floor of the chicken house with a whitewash brush. If used in this way it is sure death to lice and mites.

While the fowls are being thus easily rid of these persistent enemies, they inhale the vapors, and the throat and vital parts are being benefited at the same time.

"CARBOZIN" is **worth all it costs** as a Cholera Cure and preventative. In fact no farmer or breeder should attempt to get along without it.

In some sections fleas are the special and peculiar **pest of** poultry, some claiming they are the worst pest, and that they seriously diminish the egg product. "CARBOZIN" is as deadly to fleas as to lice and mites.

"CARBOZIN" is almost worth its weight in gold to those who seek some little profit from poultry raising, and this includes all classes, from the farmer's wife to the fancier.

We put "CARBOZIN" up in only one package, and that a three-gallon can, in case, price $2.00. Try it once and you will find your money well invested.

Send Post Office Order, Express Order, or Two Dollars in currency, and we will ship the day the order is received.

CHARLES H. CONNER & CO.,

211 Clay Street, LOUISVILLE, KY.

REMEMBER

That the prices quoted on back cover of this
book are not for culls or common stock
but for

STRICTLY FIRST-CLASS BIRDS

If you want the best you must pay the best
price.   We sometimes have special
bargains.

W. W. CLOUGH,

MEDWAY, MASS.

Keep Your Chickens

STRONG AND HEALTHY

BY USING

Sheridan's Condition Powder!

It is absolutely pure ; Highly concentrated ; Most Economical, because such small doses : No other kind one-fourth as strong ; In quantity costs less than one-tenth cent a day per hen ; No other made like it.

It gets your pullets and old hens to laying early. It prevents all disease, Cholera, Roup, Diarrhœa, Leg-Weakness, Liver Complaint, and Gapes. Large Cans are Most Economical to buy. Six, $5.00.

It is a Powerful Food Digestive.

Therefore no matter what kind of foods you are using always give with them

Sheridan's Condition Powder.

It assures perfect assimilation of the food elements, needed to maintain health, produce eggs, and assist in forming new plumage. It is worth its weight in gold when hens are moulting.

It is a fact based upon the declaration of a noble contributor to science, that through the medium of the circulating blood any particular organ of a living animal may be reached and stimulated into renewed vigor and activity if we will only administer the **proper** material to produce the desired effect.

Those who get best results from using Sheridan's Condition Powder are those who commence with little chickens, giving small doses twice a week ; then a little larger doses, and so on to time when getting the pullets ready for early fall laying. a dose say of one teaspoonful to each pint of food, and so continue to use it, as one customer says she does, namely . "from the cradle to the grave," and you will succeed nine times in ten, and have plenty of eggs to sell in the early winter when prices are highest.

If You Can't Get it Near Home, Send to Us. Ask first.

We send postpaid one pack for 25c. Five $1. A two lb. can $1 20 ; Six cans $5, express paid. Sample copy of "the best poultry paper published," sent free. One large can. price $1.20, and Farm Poultry one year. price $1.00, sent for $2.00, cash or stamps. I. S. JOHNSON & CO., 22 Custom House Street, Boston, Mass.

Clough's
BANTAM
YARDS.

The Largest Exclusive Bantam Yards in New England.

Some grand specimens in all varieties, except Game. A cordial invitation is extended to all lovers of Bantams to visit our yards. Below we quote prices, and wish to state that we ship

STRICTLY FIRST CLASS STOCK ONLY.

Prices of Eggs and Good Breeding Stock.

VARIETIES.	EGGS.		FOWLS.		
	PER 13.	PER 26.	SINGLE.	PAIRS,	TRIOS.
Buff Cochin, . .	$3,00	$6.00	$4.00	$7.00	$10.00
White Cochin, . .	3.00	6.00	4.00	7.00	10,00
Black Cochin, . .	3.00	6,00	4.00	7.00	10.00
Golden Sebright, .	3.00	5.00	4.00	7.00	10.00
Silver Sebright, . .	3.00	5.00	4.00	7.00	10.00
Rose-Combed White,	No	eggs.	4.00	7.00	10.00
Rose-Combed Black,	2.50	4.00	4.00	7.00	10.00
Black Tailed Japanese,	3.00	5.00	4.00	7.00	10.00
White Japanese, . .	3.00	5.00	4.00	7.00	10.00
Black Japanese, .	No	eggs.		No	stock.
W. C. W. Polish, .	3.00	5.00	4.00	7.00	10.00
Bearded W. C. W. Polish	5.00	8.00	5.00	10.00	15.00

Exhibition birds a matter of correspondence.

Satisfaction guaranteed in all transactions.

W. W. CLOUGH, Medway, Mass.

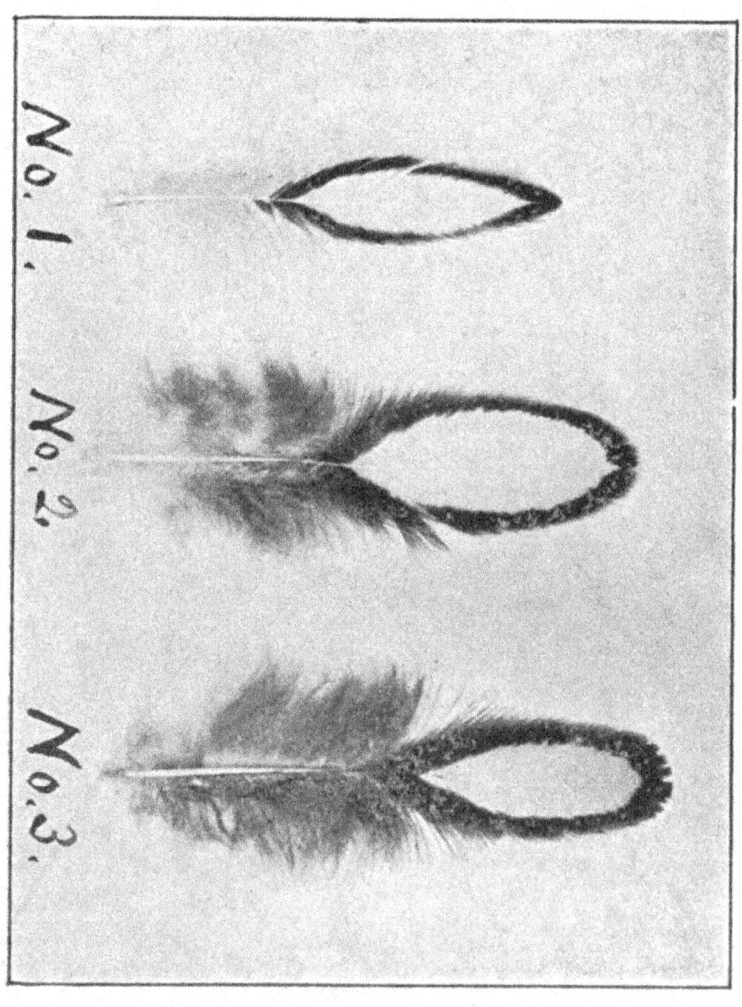

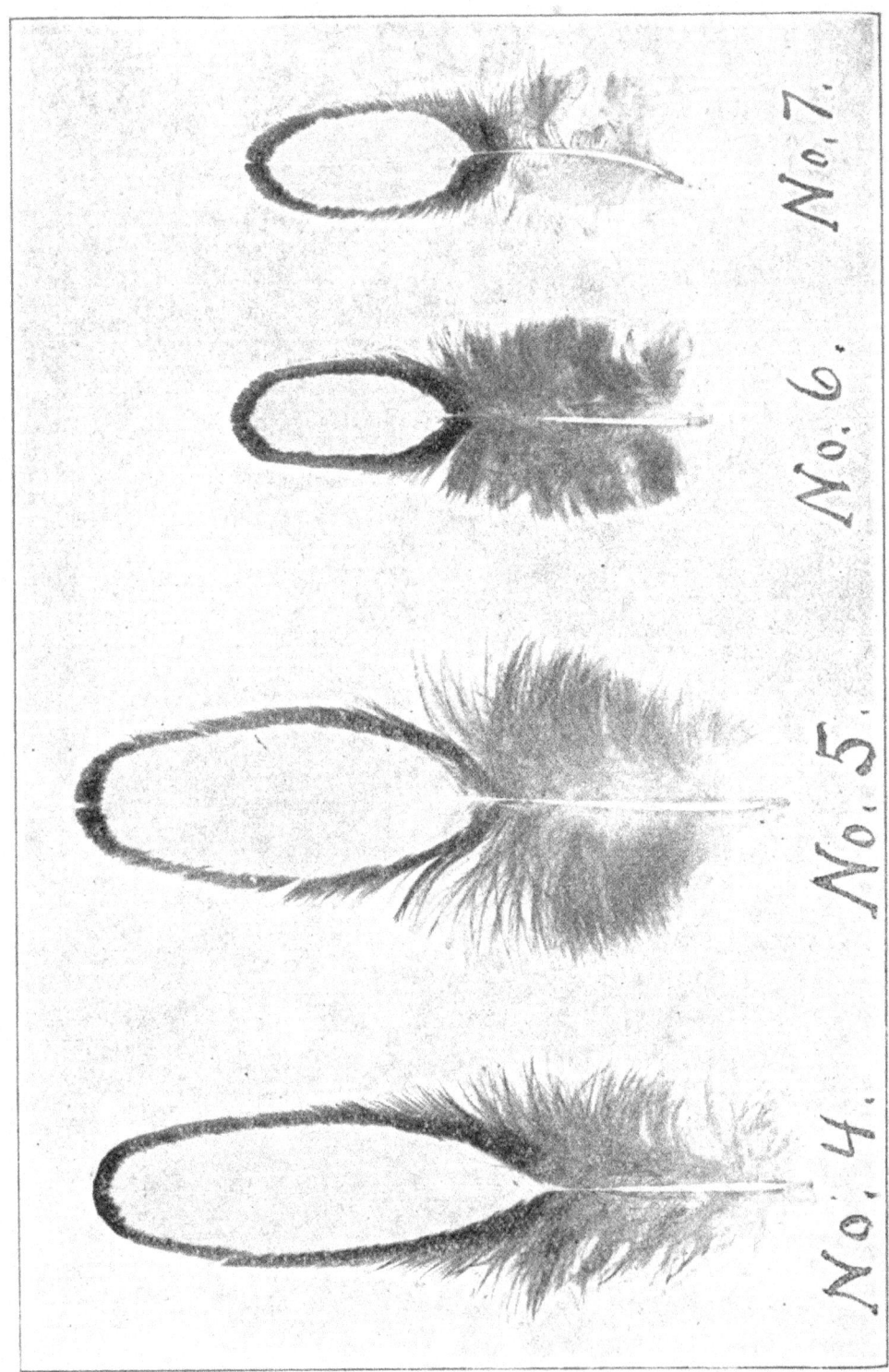